Volney Streamer

The world awheel

Volney Streamer

The world awheel

ISBN/EAN: 9783337202422

Printed in Europe, USA, Canada, Australia, Japan

Cover: Foto ©berggeist007 / pixelio.de

More available books at **www.hansebooks.com**

THE

WORLD AWHEEL

With fac-similes of water-color paintings **by**
EUGÈNE GRIVAZ

Edited by
VOLNEY STREAMER

LONDON AND NEW YORK
FREDERICK A. STOKES COMPANY
PUBLISHERS

CONTENTS.

PRELUDE.

Re-echo ye walls and ye raftered roofs,
   And voice it o'er vale and mountain,
That Pegasus, stamping his silver hoofs,
   Has opened the singing fountain.

Each drop in delight of its rhymèd might
   A song to our sport is waking,
And rainbows so bright in the soft sunlight,
   The form of the wheel are taking.

Come! mount and away!   Let the tuneful lay
   Flow on to the treadle's timing;
Yet, tarry I pray—for a moment stay
   And list to the waters rhyming.
            —S. CONANT FOSTER.

MOSCOW.

THE spires of Moscow glittering from afar
   In the pale luster of yon silver star
    Her steel clad bastions, and embattled walls,
Her domes, her fanes, and gold-bespangled halls,
No more the minstrel's midnight music hear,
No vocal strains her silent gardens cheer :—
Save where yon holy quire in pure array,
Through the gray portal treads its lonely way :
They with soft notes that sigh upon the gale,
Wake the sad echoes of the sleeping vale ;
Breathing, fair city, in a dirge to thee
Their sweetest, calmest, holiest melody.
            —Christopher Wordsworth.

A MIDWINTER REVERIE.

BEHOLD the earth enrobed as Winter's bride,
   Her snowy mantle creaks beneath the heel,
   While passing sleighs with merry music hide
The paths whereon we late did ply the wheel.

The frozen brook no longer gurgles by;
   No more the fragrant, blooming flower is seen;
The leafless tree stands naked on the sky,
   And only treasured memory is green.

No need for Milton's silent hills to speak,
   Or written log to happy hours recall;
With kindling eye and pleasure-burning cheek
   Full well, full well, we recollect them all.

Those trips awheel before the break of day,
   The pause to hear the morning songsters sing,
The break of fast on berries by the way,
   The thirst assuaged by kneeling to the spring;

The drill, the race, that memorable run,
　　Quixotic like, in search of conquests fair,—
Each 'joyed event returns like Summer sun,
　　To warm the chillness of the Winter air.

Roll on, ye frosts, and spend your rime and hoar!
　　O despot Winter, sway your substance through!
Full soon the hour when Summer reigns once more,
　　And we enjoy her ecstasies anew.

—*S. Conant Foster.*

RIVERSIDE DRIVE

NEW YORK

Painted by Eugène Grivaz

PHILLIDA ON HER WHEEL.

WHEN I was but a lad,
        Long ago,
      This simple lore I had,
  Don't you know :
That every maiden fair
Was an angel unaware,
And I wondered when and where
  The wings would grow.

But wiser now am I,
  A good deal,
Though I've sometimes seen them fly,
  Yet I feel
They are something just between
Man and angel in their mien
Since my Phillida I've seen
  On her wheel.

She does not show a sign
  Of a wing,
But her figure is divine,
  And the fling

Of her abbreviated gown,
As she flickers through the town,
Might buy the throne and crown
  Of a king !

No halo of a saint
  Does she wear,
Such as Lippo loved to paint,
  But her hair
As when all heaven streams
Through the landscape of my dreams—
In such glory floats and gleams
  On the air !

But not all for heaven she—
  Not too good !
Yet she's good enough for me
  In any mood.
And if her dashing wheel
Took her even to the de'il
Thither, too, I'd gently steal—
  Yes, I would !

                    —*Charles G. D. Roberts.*

BALLADE OF BICYCLING.

WHEN the hedgerows are sweet with bloom and bud,
  And blossoms are covering the apple-trees ;
When the air is spicy and daisies stud
  The velvety turf and emerald leas ;
When drowsily ramble the honey bees
  In perfumed nooks their sweets to steal,
And birds are rehearsing their August glees—
  Then ho for a spin on the flying wheel !

When the brain is sluggish and slow the blood
  And you're idling in knickerbockered ease,
Deep in meadow-sweet while the bitter cud
  You are ruminating of memories ;
When a pipe and a glass can not appease
  The phantoms that follow upon your heel,
And no song for sorrow will bring surcease—
  Then ho for a spin on the flying wheel !

When far to the West, where the soft clouds scud,
  A Rosalind face at the gate one sees
In the mellow light of the sunset flood,
  Her soft hair stirred by the quickening Breeze.

When darkness deepens by slow degrees
  And lily bells ring their vesper peal,
And the birds are adjourning their jubilees—
  Then ho for a spin on the flying wheel!

ENVOY.

When Winter grim takes the summer's keys
  The devil in blue a march can steal,
But when rose birds come, and the snow bird flees,
  Then ho for a spin on the flying wheel!
            —Ernest De Lancey Pierson.

THE CHAMPS ELYSEES

PARIS

Painted by Eugène Grivaz

FRANCE.

To kinder skies, where gentler manners reign,
I turn, and France displays her bright domain.
Gay, sprightly land of mirth and social ease,
Pleased with thyself whom all the world can please.
                                        —OLIVER GOLDSMITH.

MADE FOR TWO.

BY ROBERT BARR.

JACK HINKSTON was her slave, bound hand and foot to her chariot wheel, or perhaps it would be more modern to say, now that she had taken to cycling, that he was bound to her bicycle wheel. She had flouted him and scorned him for upwards of two years, and in despair Jack set himself at undoing his bonds. It was a slow and painful process, and the bonds had a habit of slipping again into hard knots when he caught a glimpse of Cissie, and Jack had almost made up his mind to emigrate to some outlandish country, for he feared he would never be his own man again unless a very broad and extensive ocean rolled between them. No matter how stern his resolutions were, they faded away to nothingness when he met Cissie in a new, pretty and stylish dress, for every costume she adopted seemed to be even more fetching than the last. Clothes have so much to do with the appearance of a pretty girl. The trouble with Jack was that he had too much respect for women in general, and they all knew that, and consequently despised him, Cissie being the leader in heaping contumely on poor Hinkston, who, after all, was a very nice sort of fellow, who did not realise that girls as a rule are somewhat silly and more apt to take up with a shallow-brained, conceited rapscallion than with a fellow of genuine worth like Jack Hinkston.

Jack had heard that Cissie had taken to the bicycle, but he had never seen the young lady on a machine. For the past week or two

Jack had avoided Cissie, and his resolutions of abandonment had so strengthened themselves that he felt he would be a free man if he could merely keep away from her, but on the other hand he realised that the next time he saw her she would have on a new dress and would look more like an angel than ever, and he trembled for the result.

In the matter of the new dress Jack was perfectly right, and he saw the girl under circumstances that nearly resulted in his downfall, but not the kind of downfall he had looked for. He had taken out his bicycle and had gone for a long wheel into the peaceful country, where he would have no comrades but the trees and the green fields and the hedges that bordered the lanes. As he cycled along a narrow country thoroughfare, wheeling at a leisurely pace, for rapid cycling doesn't lend itself to sombre meditation, he heard behind him the sharp ring of a bicycle bell. Something in its imperative clang, or else the fact that he was on a lonely road, caused him to look over his shoulder, and he nearly tumbled off his machine with amazement and surprise. There was Cissie on the top of a silver-plated machine, with the very newest and most natty cut of an advanced woman's bicycle costume, clipping over the distance at a tremendous rate of speed. She passed him with a whir, giving him a saucy nod and a salutation as she went by. Jack gasped and said under his breath, well he would be somethinged—a phrase that would not look pretty on these pages, but it must not be taken as typical of Jack's conversation. He was knocked all of a heap by the astonishing sight of Cissie in the very latest lady's bicycling costume. The next instant he put his muscles to the wheel and sped after her, shouting:

"Stop a minute, Cissie. I want to tell you something."

But the young woman never paid the slightest attention. She bent over the handle-bars and raced down that lane in a way to make pedestrians' heads swim. Jack shouted ineffectually two or three times, then pulled up and said to himself:

"Well, let her go. She will find out all I wanted to tell her."

Cissie disappeared round a corner and when Jack came to it she was not to be seen down the long avenue on which the sunshine flickered through the entwining branches of the trees overhead. Jack went on leisurely for a mile or two, then he jumped lightly off his machine and trundled it along beside him. He was now miles away from civilisation, deep in the midst of the country. The road had suddenly become very bad, and Jack, who knew the peculiarities of every lane within miles around, thought it safer to wheel the bicycle along by hand rather than risk a puncture of his pneumatic tires on the sharp flint stones scattered with such profusion along the way. Near a little rustic bridge over a clear stream at the bottom of a dell he found what he expected to find—a very pretty girl with a most woebegone, disconsolate look on her face sitting on the grassy bank looking forlornly at a bicycle that lay on the road with the tire of the hind wheel collapsed.

" Hulloa, Cissie," said Jack breezily. " Had a tumble?"

" No," snapped Cissie ; " I am not in the habit of tumbling."

" Ah," said Jack, " I see what is the matter. The tire is punctured. I knew that would happen. I shouted after you to tell you of this bit of road, but you would not listen."

" I did not hear you," said Cissie, at which assertion Jack raised his eyebrows with incredulity, which made Cissie all the more angry, especially as she knew she was telling a thing which was not.

"Well, I don't want any help from you," she said curtly.

"Why, of course not," returned Jack, sitting down on the opposite bank, and leaning his bicycle against the hedge. " A person who comes out on a wheel and doesn't know how to mend anything that goes wrong is simply a silly fool. One can see you understand all about cycling, because you have left your machine lying on the ground and the oil is running out of your lamp."

Cissie looked at the young man in amazement.

" It isn't your lamp," she said at last, "and I can surely do what I like with my own. I don't see what right you have to interfere."

"Bless you, Cissie," said Jack. "I am not interfering. I am not even offering advice. I have never yet had the pleasure of seeing a woman take off a pneumatic tire and mend the inner tube. This of course you have to do before you can move on, for you are miles away from any place, and even if you left your machine here, you would not dare to walk home in that idiotic costume."

Cissie blushed deeply and the tears came into her usually bright eyes. She tugged nervously at the skirts of her coat, and then seeing what she was doing and finding that they but scantily covered her knees she looked for a moment as if she were going to burst out crying, for it had taken some bravery to come out for the first time in knickerbockers. However, instead of crying she blazed out at him in anger:

"What business is it of yours," she cried, "how I am dressed? You are nothing to me, and I am sure I don't care a penny for your opinion one way or another.

"I don't suppose you do," said Jack, striking a match and lighting his pipe. "I used to be under the impression that you knew how to dress. I am not any longer. I used to think that you could not put on anything that would be unbecoming. Now I hold no such opinion. I once had an idea that nothing you put on would make a guy of you, but now, Cissie, that idea has fled. Still I must say that I admire your bravery in coming out in the daylight, where people can see you, in such a rig. It is utterly futile for you to pull together the skirts of that coat. The hard things you have said to me when you had on a pretty lawn tennis costume, for instance, do not affect me a bit when they are said by one who merely looks like a saucy, impudent boy. You see, Cissie, I look down upon you as you once looked down upon me."

"How dare you say you look down upon me?" said Cissie.

"Because it's true," answered the young man, calmly. "This bank is ever so much higher than the one you are sitting on, or rather were sitting on, for now you are trying to crouch out of sight, and I don't wonder at it. I take back all those numerous offers of marriage I made to you."

"You wretch!" she cried, springing to her feet. "You take them back, do you, when you know very well they were all rejected and scorned."

"Oh, Cissie, Cissie!" cried the young man, turning away his head. "Sit down again. Do sit down. The costume doesn't look so bad on a bicycle, but it is simply awful when a girl stands up."

When he looked around again Cissie had sat down and had drawn her bicycle up on its wheels, crouching in a measure behind it, as if with its spindly tires it could hide the awfulness of the costume.

"Well, Ciss," cried Jack, "when are you going to get at mending that tire?"

"I—I—I don't know anything about tires," sobbed Cissie.

"Ah," said the young man, with a long breath, "I thought that was the case. A woman never knows how to do anything well except scold. Most things in this world a man can do better than a woman, and that fact never becomes so apparent as when a woman tricks herself out as a man. Then her general futility becomes apparent, even to an infatuated fool like myself."

Cissie had bent her head upon her hands, which rested on the saddle of her cycle. It was quite evident that she was in tears, and Jack, waiting for a reply, smoked on in silence.

At last he said, in a gentler voice:

"Look here, Cissie, if you ask me very nicely I will take off that pneumatic tire and mend it in five minutes by the watch."

Cissie looked up again with something like her former indignation in her eyes.

"I'll throw the machine into the stream," she said, "before I will ask you to mend it."

"Just as you please, Cissie," replied Jack, clasping his hands behind his head and leaning back in luxurious enjoyment of his pipe. "Just as you please. The day is my own and I suppose you will wait here till night before you venture back home again.   Out of the kindness of my heart I will stay here with you, not to look at you, for I shall gaze at the tops of the trees as much as possible, and not to talk to you, for if there is anything in this world I abhor it is an impudent, cheeky boy.   But this lane is a great place for tramps and gypsies, and it becomes very dark at night, because of the overhanging trees.   It is a gruesome thoroughfare, and a nasty place in which to meet a villain after the sun has gone down."

" I have already met a villain and a brute," sobbed Cissie, who had now let the bicycle go and had buried her face in her hands.

" If you refer to me, Cissie," said Jack, " this is simply like most of the things you have said—not true.   I am only too pleased to be of any assistance to anybody, but at the same time, although you might not have thought it by my former conduct, I am too proud to offer any assistance unasked."

Jack smoked on, gazing up, as he had promised, at the tree tops. The silence was broken only by the sweet singing of the birds and now and then by a quick catch of the breath on the part of Cissie. Five minutes elapsed, and then ten.

" Jack," said Cissie, without raising her head.

" Did you speak?" inquired the young man.

" Jack," she said.   " I am perfectly helpless and I think you have been very horrid to me."

" All right," said the young man, rising to his feet.   " I will go away.   But do try to get out of this lane before darkness comes on."

" Don't go away," cried Cissie.   " Please forgive what I said, and won't you be so kind as to mend my tire?"

Jack picked up the bicycle, took off the dripping lamp, turned the machine quickly wrong side up, took the materials out of his own cycle pouch, had the tire off and on again and pumped full in an incredibly short space of time.   Righting the machine and putting the lamp on once more, he held out his hand.

Cissie reluctantly got on her feet.

"There," he said, "you see how quickly a thing is fixed when the time is not wasted in foolish conversation. Least said, soonest mended. Are you going any farther, Cissie? If you are, I would advise you to walk your machine over these stones."

"No," said Cissie, with a deep, quivering sigh, "I am going home as quickly as I can, and then I will burn this awful costume. I did not really want to put it on, but all the girls in our club have one."

"Cissie," said the young man, slipping his arm around the natty, tailor-made coat. "The costume is all right, and don't you be bluffed. It looks as pretty as a picture and suits you down to the ground. When a girl talks kindly it's simply one of the nattiest costumes that ever was constructed by a tailor, but I say, Cissie, don't you think we have misunderstood each other for a long time now, and don't you think that a bicycle made for two would require less exertion than a couple of single machines?"

"I don't know but it would," said Cissie, looking up with a smile that was all the sweeter because there was just the slightest suspicion of a quiver at the corners of her pretty lips.

And then Jack, with a villainy that surprised himself, taking advantage of the lonely situation, stooped down and kissed her, and Cissie, realising the futility of resistance, did not resist.

THE RIVIERA

Painted by Eugène Grivaz

TO MY BICYCLE.

FAR swifter than e'er Atalanta flew,
    And silent as the working of the mind
    Thou glidest, leaving city walls behind
To fly to where—in many a brilliant hue
Beneath the moon's pale light—the sparkling dew
  In trembling, scintillating drops is found ;
  Where odours sweet and fragrant fields abound
And nature breathes to man of life anew.
Amazed, I guide thee, noiseless thing of steel !
  Scarce using force to urge thee thro' the night,
Wondering if thou, like me, dost bondage feel,
  And find relief in this green-pastured flight ;
If thro' thy frame the travelled pleasures reel
  Responsive, haply, to mine own delight.

                      —*S. Conant Foster.*

THE MUSE AND THE WHEEL.

THE poet took his wheel one day
 A-wandering to go,
 But soon fell out beside the way,
The leaves allured him so.

He leaned his wheel against a tree
 And in the shade lay down ;
And more to him were bloom and bee
 Than all the busy town.

He listened to the Phœbe-bird
 And learned a thing worth knowing.
He lay so still he almost heard
 The merry grasses growing.

He lay so still he dropped asleep ;
 And then the Muse came by.
The stars were in her garment's sweep,
 But laughter in her eye.

" Poor boy !" she said, " how tired he seems !
 His vagrant feet must follow
So many loves, so many dreams,
 —To find them mostly hollow !—

" No marvel if he does not feel
    My old familiar nearness ! "
And then her gaze fell on his wheel
    And wondered at its queerness.

" Can you be Pegasus," she mused,
    " To modern mood translated,
But poorly housed, and meanly used,
    And grown attenuated ?

" Ah, no, you're quite another breed
    From him who once would follow
Across the clear Olympian mead
    The calling of Apollo !

" No Hippocrene would leap to light
    If you should stamp your hoof.
You never knew the pastures bright
    Wherein we lie aloof.

" You never drank of Helicon,
    Or strayed in Tempe's vale.
You never soared against the sun
    Till earth grew faint and pale.

" You bear my poor deluded boy
    Each latest love to see !
But Pegasus would mount with joy
    And bring him straight to me ! "

He woke.   The olden spell was strong
   Within his eager bosom ;
And so he wrote a mystic song
   Upon the nearest blossom.

He wrote, until a sudden whim
   Set all his bosom trembling ;
Then sped to woo a maiden slim
   His latest love resembling.

—Charles G. D. Roberts.

SCOTLAND

Painted by Eugène Grivaz

·

SONG.

NAE gentle dames, tho' e'er sae fair,
   Shall ever be my muse's care ;
   Their titles a' are empty show
Gie me my Highland lassie, O.

Oh were yon hills and valleys mine,
Yon palace and yon gardens fine !
The world then the love should know
I bear my Highland lassie, O.

   *   *   *   *   *   *

Altho' thro' foreign climes I range,
I know her heart will never change,
For her bosom burns with honour's glow,
My faithful Highland lassie, O.

                                       —*Robert Burns.*

TOASTING SONG.

AWAY, dull care, away !
     Till night doth bud in day,
Till dawn doth lie
In the eastern sky
With a promise bright and gay ;
     For Joy is king,
     And his subjects sing
To the wheel forever and aye.
          Away !

Uncork the wine so red
The grape of France hath bled :
     Libations pour
     Till our sorrows thaw,
And the ice of life is dead ;
     Till morning steals
     On our glistening wheels,
And the order " Mount " is said.
          Uncork !

—*S. Conant Foster.*

A CYCLE MEMORY.

INTO my thoughts half dreaming
  A picture will oft-times steal,
Of a good, straight, level highway
  And a girl on a whirling wheel.

Between the flowering hedgerows
  We are gliding and gliding along,
And our wheels are keeping together
  While my heart is singing a song.

And the song is ringing and ringing
  To the wheels' reverberant sound :
" 'Tis love that turns the world, dear
  And makes the wheels go 'round."

The hedgerows now are brown and bare,
  My love is far away,
And I have not thought of singing
  For many and many a day.

—Polly King.

HOLLAND

Painted by Eugène Grivaz

HOLLAND.

TO men of other minds my fancy flies,
    Embosom'd in the deep where Holland lies,
    Methinks her patient sons before me stand
Where the broad Ocean leans against the land,
And sedulous to stop the coming tide,
Lift the tall rampires' artificial pride.
Onward methinks and diligently slow,
The firm connected bulwark seems to grow;
Spreads its long arms amidst the watery roar,
Scoops out an empire, and usurps the shore
While the pent Ocean, rising o'er the pile,
Sees an amphibious world beneath him smile:
The slow canal, the yellow-blossom'd vale,
The willow-tufted bank, the gliding sail,
The crowded mart, the cultivated plain,—
A new creation rescued from his reign.

                —Oliver Goldsmith.

THE BICYCLE.

SPUN in some mighty wizard's brain,
　　The potent spell that gave thee birth !
　He questioned nature, not in vain,
And called thy being from the earth ;
To share the task, he summoned fire ;
　Æolus at his bidding came ;
He fashioned by his vast desire
　The mystic bond of steel and flame.

The subtle genius of the Greek,
　That bade swift Hermes tread the air,
And Icarus, on pinions weak,
　The vast ethereal spaces dare,
And Phaeton forget his fears,
　And speed the cloud-borne chariot free,—
Prophetic looked adown the years,
　And dreamt a deed fulfilled in thee.

What if he wrought not what he sung ?
　The vision into being came ;
And it were meet the Grecian tongue
　Should lend the magic wheel a name.

For sure, the god-like force that woke
   The pulsings of the Attic heart
Is present here in every spoke,
   And latent dwells in every part.

The Caliph's carpet, magic-spun,
   The Lord of Bagdad bore alone,
None other ever gazed upon
   Or mounted on that airy throne ;
The modern necromancer weaves
   A myriad mystic steeds of steel.
Alike, or king or common cleaves
   The gale upon the ready wheel.

Outdone, outdone, O genii, ye
   Who wrought that Orient fabric rare !
A nobler steed is waiting me,
   And I am regent of the air.
With regal foot I spurn the dust,
   All baser barbs are left behind,
I launch me like the lance's thrust,
   And speed triumphant down the wind.
                    —*Robert Clarkson Tongue.*

SWITZERLAND

Painted by Eugène Grivaz

SWITZERLAND.

WITHIN the Switzer's varied land,
    When Summer chases high the snow,
    You'll meet with many a youthful band
Of strangers wandering to and fro:
Through hamlet, town, and healing bath
    They haste, and rest as chance may call,
No day without its mountain path,
    No path without its waterfall.

They make the hours themselves repay,
    However well or ill be shared,
Content that they should wing their way
    Unchecked, unreckoned, uncompared:
For though the hills unshapely rise,
    And lie the colours poorly bright,
They mould them by their cheerful eyes
    And paint them with their spirits light.
                                —*Lord Houghton.*

THE BRIGANDS OF THE BRUNIG.

BY ROBERT BARR.

THERE is a saying to the effect that birds of a feather flock together; nevertheless it must be admitted that in real life members of the same profession sometimes have an unreasonable jealousy of one another. In proof of this statement I have merely to cite the well-known fact that similar cities situated near each other are generally bitter rivals; and in the same city newspapers of differing politics seldom have much love lost between them; the same, I regret to say, is true of bicycle clubs. For instance, the celebrated Fly Wheelers' Association looks down upon the Bouncing Bike Club with unconcealed contempt, while the Bouncing Bikers have no hesitation in saying that the Fly Wheelers are merely a set of conceited snobs. When the members of rival clubs meet in foreign lands one would think they should help each other rather than try to get their fellow-countrymen into trouble, but if such were the case, this pathetic account would never have been written.

As every one knows, the Fly Wheelers' costume is a subdued, almost unnoticeable grey, while the Bouncing Bikers wear a suit of black with scarlet trimmings and bright yellow stockings. The High Flyers, as the Fly Wheelers are sometimes called, look upon this costume as garish and vulgar, while their rivals maintain the others dress like a lot of maiden ladies new to bicycling. One advantage of the Bouncers' dress is that a member of the club can recognise another anywhere within two miles, and I must admit that when you meet a detachment of the Bouncers on a smooth, level road going at the rate of something less than a mile a minute, all bending over their machines and exerting themselves to the utmost, the result is something like a flash of lurid lightning gone astray. The High Flyers, on the other hand, go soberly along the road and rather resemble a gentle fleecy summer cloud as they pass by.

Now it happens that last summer two members of the High Flyers' Association in their sober grey suits were wandering around Lucerne in Switzerland, trying to find out whether it was safe to bicycle over the Brunig Pass to Interlaken. They knew that the Brunig was the lowest pass in Switzerland, and thought perhaps they might be able to wheel down it in safety and comfort, as they had also been told that on account of the new railroad from Lucerne to Brienz, there was now very little traffic across the Pass, and so they expected to find the road clear for cycling.

I am told that it is the right thing nowadays in the highest social circles to pretend ignorance on every useful subject. As long as one knows absolutely nothing and has some money and a pedigree, one is welcomed among the upper ten all over the world. I am therefore paying a high compliment to the High Flyers when I say that no body of young men I have ever met were more ignorant than they are. Their natural ignorance has been increased by the fact that each one of them is a graduate of a very swell university. Nothing develops the muscles and attenuates the brain so much as a course in one of our modern universities, and all the High Flyers are justly proud of their records. On the other hand, the Bouncing Bikers are young men who are up to snuff; they know a thing or two and actually are not ashamed of it, even if they do dress in scarlet, black and yellow.

The landlord of the hotel at which the High Flyers were stopping told them that he knew nothing at all of the Brunig Pass, but

that two young countrymen of theirs had recently come over on bicycles and would probably give them every information. The High
Flyers then sought their countrymen and found them sitting in front
of one of the lowest cafés in Lucerne, drinking plebeian beer; they
were furthermore shocked to find them in the striking costume of the
Bouncing Bikes. However, the High Flyers (for necessity knows no
law) condescended to inquire of the Bouncers how the road across
the Brunig was.

"As far as the cycling goes," said one of the Bouncers, "the
road is all right. Your machines have brakes, of course?"

"Oh, yes," was the answer.

"Very well, I should keep a touch of the brake on if I were you,
while going down, and not let the machine get out of control; still I
think that members of the High Flyer Club would better go by
train."

"Why do you say that?" inquired the spokesman of the High
Flyers.

"Because of the brigands," replied the other.

"What brigands?"

"Why, haven't you ever heard of the brigands of the Brunig?
But, of course, I don't wonder you haven't, because the Swiss people
keep very quiet about them. But surely you know why the railroad
was built?"

"No, I don't. Why was it built?"

"All on account of the brigands of the Brunig Pass. The road
never was safe, because they waylaid the diligences and robbed the
passengers."

"But couldn't the officers of the law capture them?"

"Capture them!" cried the Bouncer. "What should the law
capture them for? They're not acting against the law."

"I thought you said they robbed the passengers?"

"Oh, certainly, but that isn't against Swiss law; all the hotel-
keepers do it, and everyone else. As long as a man of Swiss nation-

ality confines his exactions to foreigners, he breaks no law. I thought you knew that in your club. Haven't you ever traveled in Switzerland before?"

"No," said the High Flyer, "this is our first visit."

"Oh, then you ought to study up the laws of a country when you come to visit it. However, unless you really are afraid, I think you can manage to take care of the brigands; at any rate, we did."

"How did you do it?"

"By out-racing them. You know, we hold the road record at home. My friend and myself simply put in our best licks on our bicycles and out-rode them, that's all."

"Out-rode them! What do you mean? You surely don't mean to intimate that the brigands are mounted on bicycles?"

"Why, of course they are. What century do you think you are living in? You fellows in the High Flyers are really too ignorant for any possible use. The brigands have all the very best machines that are in the market, and very proud they are of them. If you wouldn't mind condescending to a piece of strategy, I could tell you how you might outwit them without having to outride them, for I know you High Flyers are not much on the road."

"What would you suggest?" asked the other, coldly.

"Well, you see, the brigands are a most superstitious lot; all brigands are. Now, if for this occasion only, you wouldn't mind dressing up a bit, getting, for example, a long scarlet cloak and a peaked hat with a feather on it; putting on false moustaches, with an upward twirl about them, and cross the pass as Mephistopheles, they would then think the devil was riding down upon them. or rather, two devils, if you will both dress in the way I have suggested. The brigands will at once drop on their knees when they see you, and will not follow you until you get such a start that they cannot hope to overtake you; then, if you ride rapidly and get into Brienz you are perfectly safe. It would be rather a joke to tell of your adventure with the brigands when you get home."

One of the High Flyers demurred at the scarlet cloak, but finally they agreed to take the advice of the Bouncers, who knew the road and had been over it. In fact, the Bouncers were, as the others admitted, deucedly civil; and they all went together to a costumer's and were properly fitted out in Mephistophelian rig. The costumes were done up in a bundle, and the High Flyers took it on board the steamer with them, saying they would dress at the top of the Brunig Pass, which, as a matter of fact, they did.

When the two Bouncers saw the two High Flyers off on the steamboat they at once returned to the telegraph office and telegraphed to the president of the Interlaken Bicycle Club in the town of that name. The telegram ran:

"Two celebrated members of the High Flyers' Club of our country, perhaps the most noted club in the land, who have an extraordinary bicycle dress, resembling that of Mephistopheles, are going to visit your town. They will cycle over the Brunig Pass. Could you get as many members of your club together as possible, and meet them half way up the Pass? The High Flyers think they can beat anything on wheels that there is in this world, so the chances are they will try to race by you and beat you down into Brienz, and I send this telegram that you may be on your guard. If you can outrace them into Brienz the name of the Interlaken Club will be the most celebrated in bicycle circles in the world."

Telegraphing is cheap in Switzerland, but nevertheless this dispatch cost the boys some money, and the moment the president of the Interlaken Club received it he roused up the members to go out and beat the High Flyers. The club had entertained the two Bouncers and found them very good fellows indeed; going with them around the road from Interlaken to Brienz to the foot of the Brunig Pass, where, with a cheer, they sped their parting guests. Now, however, there was no time to cycle to Brienz, so they took the first steamer and rode from the end of the lake to the foot of the Pass; then they walked their machines about half-way up, and there the

Interlaken Club, at a bend in the road, stood their bicycles up against the wall, each wheel facing down the hill ready to be sprung upon at a moment's warning. Two scouts were sent around the bend to give the first warning of the coming of the High Flyers. The other members of the club threw themselves down on the sward at the right hand of the road going up, and there they lay and smoked, and enjoyed the beautiful view, until the scouts came running in, saying the two High Flyers were approaching in the most extraordinary costume, with cloaks flying like red banners in their rear. The boys evidently had their brakes clamped tightly on, and they kept their feet on the pedals so as to have control over their machines when coming down the steep pass. Every man now stood beside his machine ready to spring on, and presently the two High Flyers came around the bend in the road, and at once there went up a rousing cheer from the Interlaken Club. The High Flyers nearly fell off their machines in their surprise, but instantly threw off the brakes and at mighty risk of life and limb flew down the Brunig. The whole club was after them before one could say "pneumatic tire." The flaming red cloaks now stood out behind the fugitives and were glowing marks to race after. The Swiss fellows, however, were accustomed to mountain roads, and they tore down the Pass at a rate that has probably never been equaled on earth. To the horror of the High Flyers, they saw one after another glide past them, rending the air with ferocious yells of triumph. By the time the level road at the foot of the Pass was reached, two miles from Brienz, the High Flyers saw it was all up with them. Every member of the brigand crew was now between them and safety. Luckily no accident had happened to a wheel when coming down, and now the Interlaken Club simultaneously at a word sprang off their machines in the middle of the road. The High Flyers saw the jig was up, so riding towards them, each sprang off his wheel and cried "We surrender."

The Interlaken men, who were a jolly set of fellows, raised a great cheer and sprang on the two cyclists, hoisting them up in the air on

the shoulders of four stalwart Swiss men and carried them in triumph
along the road, while two of the others trundled their machines.
The Bouncers had told the High Flyers that the letters " I. B. C.,"
which they would notice on the costumes of the brigands, stood for
" International Brigand Company," of which this Brunig band were
members.

The High Flyers prepared for a horrible death, but no such fate
as that was in store for them. A great dinner had been prepared for
them at Werren's Bear Hotel on Lake Brienz, and thither the Interlaken Club conducted their prisoners. There under the influence of
excellent wine the High Flyers began gradually to take in the situation, and international amity was pledged all around.

The Bouncers told this story when they returned, but nobody
believed it. The High Flyers say, on the other hand, that they knew
all along that the Interlaken Club was to meet them, and they now
claim that the Interlaken Club holds the road record of the world
and that the High Flyers come second, being thus far ahead of the
Bouncing Bike Club.

The fact is that the truth lies neither in the statement of the
Bouncers nor the High Flyers, and now for the first time the exact
state of the case has been presented in public print.

THE RHINE

Painted by Eugène Grivaz

THE RHINE.

HILLS and towers are gazing downward
    In the mirror-gleaming Rhine,
    And my boat drives gayly onward
While the sun-rays round it shine.

Calm I watch the wavelets stealing,
    Golden gleaming, as I glide ;
Calmly too awakes the feeling
    Which within my heart I hide.

Gently greeting and assuring,
    Bright the river tempts me on ;
Well I know that face alluring !
    Death and night lie further down !

Joy above, at heart beguiling,—
    Thou'rt my own love's image, Flood !
She too knows the art of smiling,
    She can seem as calm and good.
                —Heinrich Heine.

CUPID'S WHISPERING OF THE DELIGHTS AWHEEL.

OH come, fair maid, on the flying wheel,
  Where bees hum loud and the violets grow,
And birds are singing a joyous reel
From the hills above and the vales below.
The shimmering waves, in the morning's glow,
  Laugh merrily back to the clear blue skies,
As they curl in the breezes that softly blow
  Adown from the west where the mountains rise.

The air is luscious with sweet perfumes
  That rise up out of the meadows fair,
And flow from the woods, where the creamy blooms
  Fill the trees and shrubs with richness rare.
  There isn't a shadow of sorrow or care
On the earth below or the sky above,
  But beauty and happiness everywhere,
The voice of joy and the looks of love.

There isn't a sound in the field or grove,
  There isn't a whisper the trees among,
But speaks to the spirit of peace and love,
  And sings of joy in a swinging song.
  Then come, fair maid, mount the swift flying wheel,
While all nature smiles and in beauty grows,
  And the birds their notes of gladness peal,
And the wavelets curl as the soft wind blows.

                                    —Abigail Quay.

POMPEII

Painted by Eugène Grivaz

POMPEII.

BRIGHT was the sky and blue the sea, when I
   On the paved causeway of Pompeii stood,
      Perplexed at my amazing solitude :
The silent forum, open to the sky,
The empty barracks of the soldiery,
The stone mills fixed to grind the daily food,
The houses of the rich and poorer brood,
Bath, temple, theatre, I sauntered by.
Surely, methought, the folk hath left its home
But for excursion, or high holiday ;
And soon shall I behold them swarming back,
Like busy bees that buzz about their comb
Or those gregarious birds whose airy track
Instinctive, westward, points their evening way.

—*John Bruce Norton.*

MY WHEEL AND I.

THERE'S a road we know,
 My wheel and I,
  Where we love to go,
 My wheel and I.
There the briers thick by the roadside grow,
And the fragrant birch bends its branches low,
And the cool shade tempts us to ride more slow,
 My wheel and I.

But through shade and sheen,
 My wheel and I,
By the hillsides green,
 My wheel and I,
We roll along till there's plainly seen
The bridge that crosses the deep ravine,
With its echoing rocks and the brook-laugh between,
My wheel and I,

Then's a hill we hate,
 My wheel and I ;
But we toil up straight,
 My wheel and I.
For beyond the hill is an ivy-crowned gate,
And a pair of eyes that to welcome us wait ;
If we do not haste we will surely be late,
 My wheel and I.

     —Alberto A. Bennett.

EGYPT

Painted by Eugène Grivaz

THE PYRAMIDS.

AFTER the fantasies of many a night,
    After the deep desires of many a day,
   Rejoicing as an ancient Eremite
  Upon the desert's edge at last I lay:
  Before me rose, in wonderful array,
Those works in which man has rivalled nature most
  Those Pyramids that fear no more decay
Than waves inflict upon the rockiest coast,
Or winds on mountain steeps, and like endurance boast.
                   —*Lord Houghton.*

EGYPT.

'ER Libya's hills the Day-god sinks once more,
   Brightly as when their crowns the Pharaohs wore ;
   Sweet, too, as then, red-mantled evening throws
O'er Egypt's vale the spell of rich repose ;
Soft glides and dimples 'neath the sunset smile
The stream of ruins, ancient, storied Nile :
On painted tomb, the crumbling city's site,
Falls like a shower of gold, the mellow light.
                              —*Nicholas Michell.*

WHEELS AND WHEELS.

*HE maiden with her wheel of old
   Sat by the fire to spin,
   While lightly through her careful hold
   The flax slid out and in.*

*To-day, her distaff, rock and reel
   Far out of sight are hurled,
For now the maiden with her wheel
   Goes spinning round the world.*
                          —Madeline S. Bridges.

NEWPORT

Painted by Eugène Grivaz

AWHEEL

IT'S joy to be up in the morning when the dew's on the
  grass and clover,
  And the air is full of a freshness that makes it a
   draught divine,—
To mount one's wheel and go flying away and away,—a rover
 In the wide, bright world of beauty—and all the world
  seems mine !

There's a breath of balm on the breezes from the cups of the
  wayside posies ;
 A hint of the incense-odours that blow through the hillside
  pines,
And ever a shifting landscape that some new, bright charm
  discloses
 As I flash from nooks of shadow to plains where the sunlight shines.

Along by the brambled hedges where the sweet wild roses
  redden
 In the kiss of morning sunshine that woos their leaves
  apart,

Over cool, damp sward and mosses that the sound of my
    swift flight deaden—
  I leave the world behind me and am close to Nature's
    heart.

I hear the lark in the heavens and his silver song seems
    sweeter
  Than ever before, I fancy, since I have found my wings.
Ah—the long, smooth stretch before me! and my flight grows
    blither, fleeter—
  Good bye to the lark above me who soars in the sun and
    sings!

I see a flash in the bushes, and I hear a squirrel's chatter,
  Half frightened, and full of wonder, as I go gliding by.
Perhaps—who knows?—he is saying that something strange
    is the matter
  In the world beyond the woodland, since its creatures learn
    to fly!

I am up on the windy hilltop; oh, the fair, bright world below
    me!
  I see the flash of the river through the forest at my feet.
What beauty, what strange, new beauty has Nature deigned
    to show me
  In the world of which I wearied ere I felt her warm heart
    beat!

I sing in my care-free gladness.   I am kin to the wind that's
    blowing !
  I am thrilled with the bliss of motion like the bird that
    skims the down.
I feel the blood of a gypsy in my pulses coming,—going !
  Give me my wheel for a comrade, and the king may keep
    his crown !

                        —*Eben E. Rexford.*

CENTRAL PARK

AT NIGHT

Painted by Eugène Grivaz

THE BICYCLERS.

LIKE gray moths tasting the scented world
    When the young flowers wake in June,
  They take the first breath of the summer, whirled
  To the swift wind's daring tune.

Their thin wings glide through the docile air
  And gleam at the gaudy day ;
They skim the rich earth of her odours rare,
  And silently flit away.

And when blue night sighs through her spheric dome
  For the worlds that shine afar,
Like will-o'-the-wisps they come trooping home,
  And each one bears a star.

                  —Harriet Monroe.

A PUNCTURE IN THE TIRE.

BY ROBERT BARR.

THERE is just a little bit of the devil in every bicycle. I suppose this comes through the fierce heat and the welding of things that take place before a machine is thoroughly constructed. Anyhow, there the touch of Satan is, and I know no way of eliminating it.

When we read the parable of the devil entering into the swine and the swine at once rushing to their destruction in the sea, we recognise a certain likeness to the actions of a bicycle when it gets a man at its mercy. A bike will behave beautifully on level ground, but get it on the top of a long and unknown hill and then trouble begins. I don't know whether it is because of a fellow feeling with hogs or not, but the moment I am on a bicycle going down a hill I think of the swine that rushed down to the sea.

It would be just as easy for a tire to become punctured when you are near home as when you are far away, but I never knew that to happen. The first time a man gets his pneumatic tire punctured is a very serious occasion with him. I suppose that most persons, when taking on a bicycle, have been shown by some kind friend just how easy it is to slip the pneumatic tire over the wheel rim and mend a puncture in the inner tube. In my back yard a friend took off the tire, extracted the thin inner tube (the whole operation looked like a clinical demonstration in anatomy), fixed a mythical puncture, slipped

on the outer covering again, pumped it up, and there it was. My friend
claimed that he could mend a puncture and have everything in going
order again in three minutes by the watch, and I have no doubt he
spoke the truth, but I am free to confess that I can't do it, and on
the first occasion I had to meet with a real puncture I am sure I could
not have fixed the machine inside of six weeks.

When I first took to riding the bicycle I was in constant fear of
puncturing. I dodged every piece of stone on the road, and would
not run over a piece of glass if I could help it. I seemed to be con-
stantly watching the path ahead of me for anything that would
puncture the tire, but by and by, as nothing happened, I grew more
easy in my mind, and at last completely forgot that such a thing as
punctures added to the many troubles of bicyclers.

In touring around I have a fancy for leaving the main traveled
roads and journeying along unfrequented lanes more out of the way
of traffic, relying on good luck and a reasonably accurate map to
bring me to civilisation again. It was on one of these occasions, when
I was at least twenty miles from home, as the crow flies, and right in
the depths of the country that I came just around a corner upon a
bit of new road bristling with flint stones into which I drove full
speed before I noticed what was ahead of me. The tire collapsed,
and the wheel came down with that sickening thud, thud, on the
road which, once experienced, is never forgotten. I suppose that
five or six times a year somebody comes along that lane, and among
those five or six persons there is perhaps one who never saw or heard
of a bicycle. I drew my disabled machine to the grassy side of the
lane under the hedge, and sat down with my head in my hands trying to think what that man had done first when he took off the tire.
It was months before, and though I had seemed to understand the
operation at the time my ideas now were in the most hazy condition
possible. As far as I could recollect he unscrewed something first
and then things seemed to come apart; so I attacked the valve into
which you pump the air, and finally got it into a position where I

knew I could never get it back again. I tried to slip the outer cover off, but found to my amazement that a couple of wires were concealed somewhere in the edge of the outer tire that I had entirely forgotten. After an hour's hard work I had gotten the machine in such a state that I very much doubted if it could ever be put together again. I now consulted my map to find, if possible, in what part of creation I really was. I hadn't much confidence in the map, anyhow; it had led me astray more than once, and had landed me in the most unexpected places, but now, if it told the truth, there was a main road a mile or two further ahead, and along that main road undoubtedly a bicycler would ultimately come, and the chances were that he would understand something about pneumatic tires. If I had had complete faith in the map I should have trundled my bicycle to the main road, but as it was I thought I had better mount the hill further on and see if I could discover signs of the principal thoroughfare. When I got on top of the hill I found there was another gentle decline and a still higher hill, and thus I was lured on and on, leaving my bicycle alone by the roadside until I found, as I had suspected, that there was no main road anywhere in the neighborhood. Turning back in despair, not knowing what next to do, I was overjoyed to see coming towards me, over the crest of the low hill, a man on a bicycle. When he saw me he apparently hesitated, and made as if he would turn back, but finally concluded to come on. As he approached I saw with amazement the reason of his hesitation. He was on my machine. I planted myself square in the middle of the lane, and although he attempted to dodge me, I grasped the handle bars.

"Now, you scoundrel," I said, "get off."

He was a much smaller man than I am, and I knew that if matters came to extremities I would have little difficulty in taking care of him. He got off without a word, but his face was pale and his lip trembled.

"Now," I said to him, "if I did right, I would break your ugly head. You may be thankful at getting off so easily; however, I am obliged to you for mending the tire, and because of that I will say no more about the matter."

The young fellow edged away from me and looked as if he would like to take to his heels.

"You needn't be afraid," I said ; "I won't hurt you."

Opening the little tool satchel at the back of the seat, I saw that he had not taken away the oil can and some of the other implements.

"Where are the rest of the tools?" I asked.

"They are all there that ever was there," he answered with hesitation.

It wasn't worth while discussing the matter with him, so I mounted the machine and ascended the hill he had come down and ran along where the disaster had happened. Here, to my amazement, I found my own machine still lying where I had placed it under the hedge. Without dismounting I whirled around and made after the young fellow whom I had robbed. I suppose I did look like a disreputable character with my face and clothes smeared with oil through my hours of effort at trying to take my machine to pieces. The moment the young fellow saw me coming behind him he began to run as fast as he could, and finally seeing that I was bound to overtake him he tried to get through the hedge, where he stuck.

"I have no money," he shouted, "at least only a shilling or two in my pocket."

"Look here," I said, "I am not such a villain as I look. I am no bicycle thief. I thought you were on my machine. Didn't you see it as you came along lying up against the hedge?"

"No," he answered with relief, coming down to the side of the road again.

"Well," I said, "I must apologise very abjectly for my conduct. Your machine is exactly like mine and I thought it was mine. I never expected any sane bicyclist would come along this lane."

"Didn't you come along the lane?" asked the young fellow.

"Yes, but I said sane bicyclist."

" Oh."

" I have punctured my tire, and I don't know how to repair it.
I have been working at it for hours, and if you know anything about
tires, for Heaven's sake forgive me for the robbery, and come back
and show me how to make my tire air tight once more?"

The young fellow was very nice about it indeed, and he returned
with me to where the disabled machine lay. With that expertness
which receives my most profound admiration, but which seems im-
possible for me to emulate, he whipped off the tire, mended the tube,
and put everything in shape again. The whole episode was a lesson
in Christian charity which I hope will not be lost on me.

I don't know when the second puncture of that tire will take
place, but I hope when it does I will have so good and mild a friend
within call as the young fellow whom I robbed of his bicycle.